I0475713

The Chronaseur
A Tasting Journal

Copyright© 2015 Justin Klein
All rights reserved, including the right to reproduce
this book, or portion thereof, in any form.
ISBN: 978-1-329-33995-8

The purpose of this journal is to chronicle the consumption of cannabis for individuals who like to enjoy the different flavors, aromas, and methods of consumption available. I hope this book will encourage people to savor the variation between different strains as well as the variations between the same strains from different growers and regions. This journal will be a means for you to note the flavors, appearance, method of consumption, as well as the overall experience you like the best and least. Utilize this as a quick reference guide to determine whether you would like a future choice in product, and whether that product has unbeknownst to you already been consumed by you in the method of which you are currently about to try again; because, let's be honest, sometimes we forget.

☺ Enjoy! ☺

How to Fill Out

In the dispensary section list the name and/or location from which you purchased your cannabis. This will allow you to differentiate between strains in different regions and from dispensary/growers in the same region. The date section can be used as a reference to see whether the characteristics of your cannabis, or even your palate, have changed over time. Next it is important to list the strain name of the cannabis you are tasting. Circle whether the strain is an indica, sativa, or circle both for hybrid. The $/g section is meant for you to note the difference in prices between regions, dispensaries, and strains.

The next few sections are meant to help in describing the cannabis prior to consumption. Utilize the blank bars under each heading to shade in the level of each descriptor written beneath each bar. The blank description area of this section should be used to elaborate on everything that the shaded bars are not fully expressing, as well as, whether this is an extract, concentrate, or edible.

These next sections will be used to describe the consumption of the cannabis. Under the method section, circle whether you consumed the cannabis through a joint, waterpipe, pipe, or edible. Fill out the taste and smoke sections in the same manner as you did with the bar graphs above. The blank description under this section should be utilized to elaborate on the type of method of consumption, as well as the characteristics of the taste and smoke.

The final section is to capture your overall impression of the cannabis strain. Rate your overall experience on a scale of 1 through 10 and circle yes or no depending on whether you would recommend this strain to a peer.

Under experience you can describe what type of affect the cannabis had on you and a story of something that may have occurred while under the influence of that particular strain, using that particular method.

The following three pages are examples of how to complete the pages of your journal.

Dispensary (name/location) __XYZ Naturals - MT_____ date_7/15_

Strain_ _Silvertip_____ (indica/sativa) ____10_ $/g

Color

Dark

Light

Green Orange Purple

Crystal

Large — Heavy

Small — Light

Size Coverage

Density

Rock — Tight   Heavy

Loose — Airy   Light

Scent

Heavy

Light

Pine/Earthy Citrus Skunk/Cheese Floral

Description__ _Semi-dense bud, easy to break up, lots of crystals,_
_Full bodied scent with dominate notes of skunk and floral_____

Taste

Heavy

Light

Pine/Earthy Citrus Skunk/Cheese Floral

Smoke

Heavy — Full

Light — Whispy

Description_____ _Smoked out of blue Sherlock named "Petey"_____
_____ _heavier smoke, with dominate earthy/florally flavor_____

Overall ____8_/10   Recommend (Yes)/ No

Experience____ _very heady/relaxing high, great for nausea, movies,_
_passenger on road trips, and blowing smoke rings_____

Dispensary (name/location) _Chalice – San Bernardino, CA_ date _7/15_

Strain _Blue Dream_   indica/(sativa)   _35_ $/g

Color	Crystal	Density	Scent

Color — Dark / Light — Green, Orange (crossed out), Purple, Amber

Crystal — Large / Small (Size), Heavy / Light (Coverage (crossed out) — Clarity)

Density — Rock / Loose, Tight / Airy, Heavy / Light

Scent — Heavy / Light — Pine/Earthy, Citrus, Skunk/Cheese, Floral

Description _Light amber/gold shatter. See through in light_
Maintained the scent properties of its flower.

Taste — Heavy / Light — Pine/Earthy, Citrus, Skunk/Cheese, Floral

Smoke — Heavy / Light, Full / Whispy

Description _Green slyme mothership with quartz bucket swing._
clean burn, light airy smoke, floral with light citrus mix

Overall _____ _6_/10   Recommend (Yes) No

Experience _lighter focusing high, recommend for daytime use_

Dispensary (name/location) _ABC Alt. Medicine - WA_ date _7/15_

Strain _75mg Raspberry Crumble - shatter_ indica/sativa _10_ $/g

Color
Green | Orange | Purple

Dark — Light

Crystal
Large — Small (Size)
Heavy — Light (Coverage)

Density
Rock — Loose (Tight)
Heavy — Light
Tight — Airy

Scent
Pine/Earthy | Citrus | Skunk/Cheese | Floral

Heavy — Light

Description _Graham cracker crust with raspberry jelly_
topped with crumbles – no noticeable cannabis smell

Taste
Pine/Earthy | Citrus | Skunk/Cheese | Floral

Heavy — Light

Smoke
Heavy — Light
Full — Whispy

Description _Delicious! Made with real raspberries_
slight cannabis aftertaste

Overall _7_/10   Recommend (Yes)/ No

Experience _mellow, enjoyable high. Body high really_
kicked in after smoking

Dispensary (name/location)_____date___

Strain_____ indica/sativa _____$/g

Color	Crystal	Density	Scent

Color
Dark
Light
Green Orange Purple

Crystal
Large Heavy
Small Light
Size Coverage

Density
Rock Tight Heavy
Loose Airy Light

Scent
Pine/Earthy Citrus Skunk/Cheese Floral

Description_____
_____

Taste
Heavy
Light
Pine/Earthy Citrus Skunk/Cheese Floral

Smoke
Heavy Full
Light Whispy

Description_____
_____

Overall _____/10   Recommend   Yes / No

Experience_____
_____
_____

Dispensary (name/location)_____date___

Strain_____ indica/sativa _____$/g

Color
Dark

Light

Green   Orange   Purple

Crystal
Large   Heavy

Small   Light

Size   Coverage

Density
Rock   Tight

Loose   Airy

Scent
Heavy

Light

Pine/Earthy   Citrus   Skunk/Cheese   Floral

Description_____
_____

Taste
Heavy

Light

Pine/Earthy   Citrus   Skunk/Cheese   Floral

Smoke
Heavy   Full

Light   Whispy

Description_____
_____

Overall _____/10   Recommend   Yes / No

Experience_____
_____
_____

Dispensary (name/location)_____date___

Strain_____ indica/sativa _____$/g

Color
Dark

Light

Green   Orange   Purple

Crystal
Large   Heavy

Small   Light

Size   Coverage

Density
Rock   Tight   Heavy

Loose   Airy   Light

Scent

Pine/Earthy   Citrus   Skunk/Cheese   Floral

Description_____

_____

Taste
Heavy

Light

Pine/Earthy   Citrus   Skunk/Cheese   Floral

Smoke
Heavy   Full

Light   Whispy

Description_____

_____

Overall _____/10   Recommend   Yes / No

Experience_____

_____

_____

Dispensary (name/location)_____date___

Strain_____ indica/sativa _____$/g

Color | Crystal | Density | Scent

Color
Dark
Light
Green Orange Purple

Crystal
Large Heavy
Small Light
Size Coverage

Density
Rock Tight
Loose Airy

Scent
Heavy
Light
Pine/Earthy Citrus Skunk/Cheese Floral

Description_____
_____

Taste | Smoke

Taste
Heavy
Light
Pine/Earthy Citrus Skunk/Cheese Floral

Smoke
Heavy Full
Light Whispy

Description_____
_____

Overall _____/10   Recommend   Yes / No

Experience_____
_____
_____

Dispensary (name/location)_____date___

Strain_____ indica/sativa _____$/g

Color

Dark

Light

Green Orange Purple

Crystal

Large Heavy

Small Light

Size Coverage

Density

Rock Tight

Loose Airy

Scent

Heavy

Light

Pine/Earthy Citrus Skunk/Cheese Floral

Description_____
_____

Taste

Heavy

Light

Pine/Earthy Citrus Skunk/Cheese Floral

Smoke

Heavy Full

Light Whispy

Description_____
_____

Overall _____/10   Recommend   Yes / No

Experience_____
_____
_____

Dispensary (name/location)_____date___

Strain_____ indica/sativa _____$/g

Color

Green Orange Purple

Dark
Light

Crystal

Size Coverage

Large
Small

Heavy
Light

Density

Rock
Loose

Tight
Airy

Scent

Heavy
Light

Pine/Earthy Citrus Skunk/Cheese Floral

Description_____
_____

Taste

Heavy
Light

Pine/Earthy Citrus Skunk/Cheese Floral

Smoke

Heavy
Light

Full
Whispy

Description_____
_____

Overall _____/10   Recommend   Yes / No

Experience_____
_____
_____

Dispensary (name/location)_____date___

Strain_____ indica/sativa _____$/g

Color

Green	Orange	Purple

Dark / Light

Crystal

Large / Small (Size)

Heavy / Light (Coverage)

Density

Rock / Loose

Tight / Airy

Scent

Heavy / Light

Pine/Earthy	Citrus	Skunk/Cheese	Floral

Description_____
_____

Taste

Heavy / Light

Pine/Earthy	Citrus	Skunk/Cheese	Floral

Smoke

Heavy / Light

Full / Whispy

Description_____
_____

Overall _____/10   Recommend   Yes / No

Experience_____
_____
_____

Dispensary (name/location)_____date___

Strain_____ indica/sativa _____$/g

Color
Dark

Light

Green Orange Purple

Crystal
Large Heavy

Small Light

Size Coverage

Density
Rock Tight

Loose Airy

Scent
Heavy

Light

Pine/Earthy Citrus Skunk/Cheese Floral

Description_____
_____

Taste
Heavy

Light

Pine/Earthy Citrus Skunk/Cheese Floral

Smoke
Heavy Full

Light Whispy

Description_____
_____

Overall _____/10   Recommend   Yes / No

Experience_____
_____
_____

Dispensary (name/location)_____date___

Strain_____ indica/sativa _____$/g

Color   Crystal   Density   Scent

Dark   Large   Heavy   Rock   Tight Heavy

Light   Small   Light   Loose Airy   Light

Green Orange Purple   Size Coverage   Pine/Earthy Citrus Skunk/Cheese Floral

Description_____
_____

                              Taste   Smoke

Heavy   Heavy   Full

Light   Light   Whispy

Pine/Earthy Citrus Skunk/Cheese Floral

Description_____
_____

Overall _____/10   Recommend   Yes / No

Experience_____
_____
_____

Dispensary (name/location)_____date___

Strain_____ indica/sativa _____$/g

Color
Dark

Light

Green Orange Purple

Crystal
Large   Heavy

Small   Light

Size Coverage

Density
Rock   Tight

Loose   Airy

Scent
Heavy

Light

Pine/Earthy Citrus Skunk/Cheese Floral

Description_____
_____

Taste
Heavy

Light

Pine/Earthy Citrus Skunk/Cheese Floral

Smoke
Heavy   Full

Light   Whispy

Description_____
_____

Overall _____/10

Recommend   Yes / No

Experience_____
_____
_____

18

Dispensary (name/location)_____date___

Strain_____ indica/sativa _____$/g

Color

Dark

Light

Green

Orange

Purple

Crystal

Large — Heavy

Small — Light

Size

Coverage

Density

Rock — Tight

Loose — Airy

Scent

Heavy

Light

Pine/Earthy

Citrus

Skunk/Cheese

Floral

Description_____

_____

Taste

Heavy

Light

Pine/Earthy

Citrus

Skunk/Cheese

Floral

Smoke

Heavy — Full

Light — Whispy

Description_____

_____

Overall _____/10   Recommend   Yes / No

Experience_____

_____

_____

Dispensary (name/location)_____date___

Strain_____ indica/sativa _____$/g

Color

Green Orange Purple

Dark

Light

Crystal

Large

Small

Heavy

Light

Size Coverage

Density

Rock

Loose

Tight

Airy

Scent

Heavy

Light

Pine/Earthy Citrus Skunk/Cheese Floral

Description_____
_____

Taste

Heavy

Light

Pine/Earthy Citrus Skunk/Cheese Floral

Smoke

Heavy

Light

Full

Whispy

Description_____
_____

Overall _____/10   Recommend   Yes / No

Experience_____
_____
_____

Dispensary (name/location)_____date___

Strain_____ indica/sativa _____$/g

Color	Crystal	Density	Scent

Color
Dark
Light

Green   Orange   Purple

Crystal
Large   Heavy
Small   Light

Size   Coverage

Density
Rock   Tight
Loose   Airy

Scent
Heavy
Light

Pine/Earthy   Citrus   Skunk/Cheese   Floral

Description_____
_____

Taste
Heavy
Light

Pine/Earthy   Citrus   Skunk/Cheese   Floral

Smoke
Heavy   Full
Light   Whispy

Description_____
_____

Overall _____/10   Recommend   Yes / No

Experience_____
_____
_____

Dispensary (name/location)_____date___

Strain_____ indica/sativa _____$/g

Color
Green Orange Purple

Dark
Light

Crystal
Large — Heavy
Small — Light

Size Coverage

Density
Rock — Tight
Loose — Airy

Scent
Heavy
Light

Pine/Earthy Citrus Skunk/Cheese Floral

Description_____
_____

Taste
Heavy
Light

Pine/Earthy Citrus Skunk/Cheese Floral

Smoke
Heavy — Full
Light — Whispy

Description_____
_____

Overall _____/10   Recommend   Yes / No

Experience_____
_____
_____

Dispensary (name/location)_____date___

Strain_____ indica/sativa _____$/g

Color

Dark

Light

Green Orange Purple

Crystal

Large — Heavy

Small — Light

Size Coverage

Density

Rock — Tight   Heavy

Loose — Airy   Light

Scent

Pine/Earthy Citrus Skunk/Cheese Floral

Description_____
_____

Taste

Heavy

Light

Pine/Earthy Citrus Skunk/Cheese Floral

Smoke

Heavy — Full

Light   Whispy

Description_____
_____

Overall _____/10   Recommend   Yes / No

Experience_____
_____
_____

Dispensary (name/location)_____date___

Strain_____ indica/sativa _____$/g

Color
Green | Orange | Purple
Dark
Light

Crystal
Large — Heavy
Small — Light
Size | Coverage

Density
Rock — Tight
Loose — Airy

Scent
Heavy
Light
Pine/Earthy | Citrus | Skunk/Cheese | Floral

Description_____
_____

Taste
Heavy
Light
Pine/Earthy | Citrus | Skunk/Cheese | Floral

Smoke
Heavy — Full
Light — Whispy

Description_____
_____

Overall _____/10   Recommend   Yes / No

Experience_____
_____
_____

Dispensary (name/location)_____date___

Strain_____ indica/sativa _____$/g

Color	Crystal	Density	Scent

Color: Dark / Light — Green, Orange, Purple

Crystal: Large / Small (Size), Heavy / Light (Coverage)

Density: Rock / Loose, Tight / Airy, Heavy / Light

Scent: Pine/Earthy, Citrus, Skunk/Cheese, Floral

Description_____
_____

Taste   Smoke

Taste: Heavy / Light — Pine/Earthy, Citrus, Skunk/Cheese, Floral

Smoke: Heavy / Light, Full / Whispy

Description_____
_____

Overall _____/10   Recommend   Yes / No

Experience_____
_____
_____

25

Dispensary (name/location)_____date___

Strain_____ indica/sativa _____$/g

Color	Crystal	Density	Scent

Color

Dark

Light

Green Orange Purple

Crystal

Large   Heavy

Small   Light

Size Coverage

Density

Rock   Tight

Loose   Airy

Scent

Heavy

Light

Pine/Earthy Citrus Skunk/Cheese Floral

Description_____
_____

Taste

Heavy

Light

Pine/Earthy Citrus Skunk/Cheese Floral

Smoke

Heavy   Full

Light   Whispy

Description_____
_____

Overall _____/10   Recommend   Yes / No

Experience_____
_____
_____

Dispensary (name/location)_____date___

Strain_____ indica/sativa _____$/g

Color
Dark

Light

Green Orange Purple

Crystal
Large Heavy

Small Light

Size Coverage

Density
Rock Tight

Loose Airy

Scent
Heavy

Light

Pine/Earthy Citrus Skunk/Cheese Floral

Description_____

_____

Taste
Heavy

Light

Pine/Earthy Citrus Skunk/Cheese Floral

Smoke
Heavy Full

Light Whispy

Description_____

_____

Overall _____/10   Recommend   Yes / No

Experience_____

_____

_____

Dispensary (name/location)_____date___

Strain_____ indica/sativa _____$/g

Color

Dark

Light

Green   Orange   Purple

Crystal

Large   Heavy

Small   Light

Size   Coverage

Density

Rock   Tight

Loose   Airy

Scent

Heavy

Light

Pine/Earthy   Citrus   Skunk/Cheese   Floral

Description_____
_____

Taste

Heavy

Light

Pine/Earthy   Citrus   Skunk/Cheese   Floral

Smoke

Heavy   Full

Light   Whispy

Description_____
_____

Overall _____/10   Recommend   Yes / No

Experience_____
_____
_____

Dispensary (name/location)_____date___

Strain_____ indica/sativa _____$/g

Color	Crystal	Density	Scent

Color
Dark
Light
Green Orange Purple

Crystal
Large Heavy
Small Light
Size Coverage

Density
Rock Tight Heavy
Loose Airy Light

Scent
Pine/Earthy Citrus Skunk/Cheese Floral

Description_____
_____

Taste
Heavy
Light
Pine/Earthy Citrus Skunk/Cheese Floral

Smoke
Heavy Full
Light Whispy

Description_____
_____

Overall _____/10   Recommend   Yes / No

Experience_____
_____
_____

Dispensary (name/location)_____date___

Strain_____ indica/sativa _____$/g

Color
Dark

Light

Green Orange Purple

Crystal
Large Heavy

Small Light

Size Coverage

Density
Rock Tight

Loose Airy

Scent
Heavy

Light

Pine/Earthy Citrus Skunk/Cheese Floral

Description_____
_____

Taste
Heavy

Light

Pine/Earthy Citrus Skunk/Cheese Floral

Smoke
Heavy Full

Light Whispy

Description_____
_____

Overall _____/10   Recommend   Yes / No

Experience_____
_____
_____

Dispensary (name/location)_____date___

Strain_____ indica/sativa _____$/g

Color   Crystal   Density   Scent

Color:
- Green
- Orange
- Purple
- Dark / Light

Crystal:
- Size — Large / Small
- Coverage — Heavy / Light

Density:
- Rock / Loose
- Tight / Airy

Scent:
- Heavy / Light
- Pine/Earthy
- Citrus
- Skunk/Cheese
- Floral

Description_____
_____

Taste   Smoke

Taste:
- Heavy / Light
- Pine/Earthy
- Citrus
- Skunk/Cheese
- Floral

Smoke:
- Heavy / Light
- Full / Whispy

Description_____
_____

Overall _____/10   Recommend   Yes / No

Experience_____
_____
_____

Dispensary (name/location)_____date___

Strain_____ indica/sativa _____$/g

Color

Green Orange Purple
Dark
Light

Crystal

Large Heavy
Small Light
Size Coverage

Density

Rock Tight
Loose Airy

Scent

Heavy
Light
Pine/Earthy Citrus Skunk/Cheese Floral

Description_____
_____

Taste

Heavy
Light
Pine/Earthy Citrus Skunk/Cheese Floral

Smoke

Heavy Full
Light Whispy

Description_____
_____

Overall _____/10   Recommend   Yes / No

Experience_____
_____
_____

Dispensary (name/location)_____date___

Strain_____ indica/sativa _____$/g

Color	Crystal	Density	Scent

Color — Dark / Light — Green, Orange, Purple

Crystal — Large/Small (Size), Heavy/Light (Coverage)

Density — Rock/Loose, Tight/Airy

Scent — Heavy/Light — Pine/Earthy, Citrus, Skunk/Cheese, Floral

Description_____

_____

Taste — Heavy/Light — Pine/Earthy, Citrus, Skunk/Cheese, Floral

Smoke — Heavy/Light, Full/Whispy

Description_____

_____

Overall _____/10   Recommend   Yes / No

Experience_____

_____

_____

Dispensary (name/location)_____date___

Strain_____ indica/sativa _____$/g

Color
Dark
Light

Green Orange Purple

Crystal
Large Heavy
Small Light

Size Coverage

Density
Rock Tight
Loose Airy

Scent
Heavy
Light

Pine/Earthy Citrus Skunk/Cheese Floral

Description_____
_____

Taste
Heavy
Light

Pine/Earthy Citrus Skunk/Cheese Floral

Smoke
Heavy Full
Light Whispy

Description_____
_____

Overall _____/10   Recommend   Yes / No

Experience_____
_____
_____

Dispensary (name/location)_____date___

Strain_____ indica/sativa _____$/g

Color	Crystal	Density	Scent

Color
Dark
Light

Green Orange Purple

Crystal
Large Heavy
Small Light

Size Coverage

Density
Rock Tight Heavy
Loose Airy Light

Scent

Pine/Earthy Citrus Skunk/Cheese Floral

Description_____
_____

Taste
Heavy
Light

Pine/Earthy Citrus Skunk/Cheese Floral

Smoke
Heavy Full
Light Whispy

Description_____
_____

Overall _____/10   Recommend   Yes / No

Experience_____
_____
_____

Dispensary (name/location)_____date___

Strain_____ indica/sativa _____$/g

Color   Crystal   Density   Scent

Dark   Large   Heavy   Rock   Tight   Heavy

Light   Small   Light   Loose   Airy   Light

Green Orange Purple   Size Coverage   Pine/Earthy Citrus Skunk/Cheese Floral

Description_____
_____

Taste   Smoke

Heavy   Heavy   Full

Light   Light   Whispy

Pine/Earthy Citrus Skunk/Cheese Floral

Description_____
_____

Overall _____/10   Recommend   Yes / No

Experience_____
_____
_____

Dispensary (name/location)_____date___

Strain_____ indica/sativa _____$/g

Color
Green | Orange | Purple

Dark
Light

Crystal
Large | Heavy
Small | Light

Size | Coverage

Density
Rock | Tight
Loose | Airy

Scent
Heavy
Light

Pine/Earthy | Citrus | Skunk/Cheese | Floral

Description_____

_____

Taste
Heavy
Light

Pine/Earthy | Citrus | Skunk/Cheese | Floral

Smoke
Heavy | Full
Light | Whispy

Description_____

_____

Overall _____/10   Recommend   Yes / No

Experience_____

_____

_____

Dispensary (name/location)_____date___

Strain_____ indica/sativa _____$/g

Color	Crystal	Density	Scent

Color

Dark

Light

Green　Orange　Purple

Crystal

Large　Heavy

Small　Light

Size　Coverage

Density

Rock　Tight

Loose　Airy

Scent

Heavy

Light

Pine/Earthy　Citrus　Skunk/Cheese　Floral

Description_____
_____

Taste

Heavy

Light

Pine/Earthy　Citrus　Skunk/Cheese　Floral

Smoke

Heavy　Full

Light　Whispy

Description_____
_____

Overall _____/10   Recommend   Yes / No

Experience_____
_____
_____

Dispensary (name/location)_____date___

Strain_____ indica/sativa _____$/g

Color

Dark

Light

Green
Orange
Purple

Crystal

Large   Heavy

Small   Light

Size
Coverage

Density

Rock   Tight

Loose   Airy

Scent

Heavy

Light

Pine/Earthy
Citrus
Skunk/Cheese
Floral

Description_____
_____

Taste

Heavy

Light

Pine/Earthy
Citrus
Skunk/Cheese
Floral

Smoke

Heavy   Full

Light   Whispy

Description_____
_____

Overall _____/10   Recommend   Yes / No

Experience_____
_____
_____

Dispensary (name/location)_____date___

Strain_____ indica/sativa _____$/g

Color
Dark
Light

Green Orange Purple

Crystal
Large Heavy
Small Light

Size Coverage

Density
Rock Tight
Loose Airy

Scent
Heavy
Light

Pine/Earthy Citrus Skunk/Cheese Floral

Description_____
_____

Taste
Heavy
Light

Pine/Earthy Citrus Skunk/Cheese Floral

Smoke
Heavy Full
Light Whispy

Description_____
_____

Overall _____/10   Recommend   Yes / No

Experience_____
_____
_____

I hope you enjoyed "working" your way through

this journal. May this be an addition to many volumes

of The Chronaseur in your collection!

www.ingramcontent.com/pod-product-compliance
Lightning Source LLC
Chambersburg PA
CBHW021940170526
45157CB00005B/2372